P9-BIZ-863

THE POWER OF TREES

TEXT BY GRETCHEN C. DAILY

PHOTOGRAPHS BY CHARLES J. KATZ JR.

THE POWER OF Trees

TRINITY UNIVERSITY PRESS   SAN ANTONIO

In honor of Charles D. Daily, M.D.

who lived with courage, dignity, kindness, and grace

and Charles J. Katz

a loving father and my mentor in photography

When you think of a tree, you tend to think of a distinctly defined object; and on a certain level . . . it is. But when you look more closely . . . you will see that ultimately it has no independent existence. When you contemplate it, you will find that it dissolves into an extremely subtle net of relationships that stretches across the universe. The rain that falls on its leaves, the wind that sways it, the soil that nourishes and sustains it, all the seasons and the weather, moonlight and starlight and sunlight—all form part of this tree. As you begin to think about the tree more and more, you will discover that everything in the universe helps to make the tree what it is; that it cannot at any moment be isolated from anything else; and that at every moment its nature is subtly changing.

— SOGYAL RINPOCHE

Trees seem so still . . .

Trees seem so still, yet they are among the greatest life forces of movement on Earth. They draw their power across 93 million miles, from the sun. They use it to drive cycles of six fundamental elements—carbon, hydrogen, oxygen, nitrogen, phosphorus, and sulfur— that together make up 95 percent of all living tissue in the biosphere, the thin layer of life at Earth's surface. These cycles are at play when, for example, a person eats shrimp that were nourished amid the roots of a mangrove forest: the shrimp and then the person take up fundamental elements from the mangroves.

One could say that the essence of life is renewal. And in material terms, renewal comes about through the chemical processes of metabolism, most dramatically when a life comes to a close. Then the physical body containing it is transformed, consumed sometimes by fire and mostly by other living beings, from microbes to lions to giant sequoias. Fundamentally, we are indeed all of the same cloth, and the very constituents of our own bodies have likely resided in forests for many periods of time.

Through fine roots, trees draw water from underground, some from depths of over 200 feet. A single large tree may pump over a ton of water into the sky in a day, making no sound or visible sign. Two hundred such trees can produce a cumulus cloud. And a whole forest can give the sky feeling. In the Amazon, about 50 percent of falling rain is first made into clouds by trees. All told, the forests of the Amazon contribute about 8 trillion tons of water to the atmosphere each year, creating "rivers in the sky" that are a powerful influence on atmospheric circulation globally.

. the essence of life is renewal

Leaves are nature's most sophisticated solar energy systems. Through delicate pores, they take up carbon dioxide from the air and cleave it into carbon and oxygen, releasing the oxygen that we and all other animals need to breathe. Carbon dioxide has been measured from the top of Mauna Loa in Hawai'i since 1958; its concentration rises and falls rhythmically with the seasons, driven by forests in the Northern hemisphere where Earth's land mass is concentrated. Lush new growth soaks up carbon dioxide in the northern spring and returns it to the atmosphere in the autumn fall and decay of leaves.

Leaves are also masters of chemistry, combining the carbon with hydrogen (from water) and other fundamental elements to yield a complex array of sugars, fats, and proteins that nourish the plant and then feed our animal world.

These substances serve trees in many ways, for example as pigments for capturing light; as colors and fragrances for luring insects to flowers or for attracting seed-dispersing birds and bats to fruits; and as highly toxic defensive chemicals to repel hungry animals from eating the body parts a tree wants to protect. These substances serve humanity in many ways too, not least in our pharmaceuticals, most of which trace their origin to nature's masters of chemistry.

Some trees clearly can talk to one another, though no one knows how widespread this ability is. When attacked by insects or other predators, certain trees emit airborne chemicals that signal trouble to downwind trees, which in turn boost their own production of chemical defenses.

Some trees can talk to one another . . .

Trees come in at least 60,000 varieties. Over their some 400-million-year history, they have staked their claim in nearly every terrestrial environment. They stand alone on windswept expanses of sand and jut out of craggy cliff faces. They live in great togetherness, canopies tightly interlinked. In seasonally flooded parts of the Amazon, under high water, fish swim among their branches and pluck their fruits.

After trees, Earth waited 165 million years before the first mammals appeared. And another 145 million years before the first monkeylike creatures swung from branch to branch in treetops. In their arboreal home, our early ancestors acquired three-dimensional stereo vision and reliance on sight much more than sense of smell.

. . . trees depend on intimate partnerships . . .

Today trees depend on intimate partnerships with others—for some, on birds to open their cones and deliver the next generation; for others, on tiny wasps to pierce their figs and pollinate flowers hidden within; and for most, on fungi to supply nutrition to their roots. In the Neotropics about 85 percent of tree species depend on birds and mammals for dispersing seeds—sending them off to hopefully favorable sites for germination and growth.

. . . trees became trees several times over

Across Earth's land surface, wherever there was enough moisture and warmth, trees came to be the dominant organisms, shading out their competitors. When forms of algae, from which trees eventually evolved, first moved onto land 450 million years ago, their biggest challenges were gravity and thirst. In fact, the earliest plants—mosses, liverworts, and hornworts—to this day remain tiny and restricted to damp places.

Trees escaped these constraints with two revolutionary adaptations: vessels for conducting water up against the force of gravity, and lignin for making the hard and rigid trunk that supports these vessels. Palms, conifers, and flowering trees evolved these same adaptations independently at various times, so that trees became trees several times over.

How far can trees push the limits?

Water is drawn up from roots to leaves through columns of dead cells, called xylem. An average trunk has hundreds of millions of these interconnected vessels that together span a few thousand miles in length. As trees grow taller and taller, gravity and the sheer length of the pipes make it difficult to pull water to the very top leaves. Thus the ultimate height seems to be limited by water supply.

The history of trees is a story of growing taller and taller, in a very slow motion race against other trees competing for light. How far can trees push the limits? The tallest known tree on Earth is a coast redwood (*Sequoia sempervirens*) that towers 379.1 feet above the forest floor in Northern California. The maximum tree height is thought to be around 425 feet, about the height of the tallest recorded trees of the past. Today over 95 percent of the original old growth forest has been logged, likely taking with it many ancient trees taller than the giants still standing.

Trees make tricky decisions every day. They tell their living layer just how many wood cells to make to support the leafy crown and how much bark to build to protect the wood beneath it. The outer bark is the tree's protective shield. It allows trees to live through almost anything—intense rain and drought, heat and cold, insect attack, even fire.

Heartwood is the central supporting pillar of
the tree, and in some species it is as strong as
steel. It will not decay or lose strength while
the tree's outer layers of bark are intact. All
of the growth in a tree trunk occurs in a thin
inner layer, just outside the inner heartwood.
This layer grows in response to hormones—
produced by leaf buds at the ends of branches—that reflect a tree's sense of the world.
They make the tree reach up, toward light.

Heartwood is . . . as strong as steel

We tend to think of trees as having one trunk, and most do. But many species have flexible architecture, growing as light, multistemmed shrubs in arid lowlands and as majestic trees with single trunks in nearby mountains. The record in trunks goes to the Great Banyan Tree, a single individual in Calcutta with about 3,000 trunks, all connected aboveground and spanning an area of 1.5 acres. The original trunk was removed in 1925, but the tree keeps growing, sending down new prop roots to support its sprawling network of branches.

Aspen colonies exhibit more subtle architectural wonders. One colony in Utah, called Pando, comprises about 47,000 trees—genetically identical stems connected underground by a massive root system. Pando is thought to be the heaviest organism on Earth, weighing an estimated 7,275 tons and spanning about 100 acres. The whole colony appears to be at least 80,000 years old and possibly more than a million years old.

. . . the tree keeps growing . . .

Tree rings let us

Trees are the longest lived organisms on Earth. Among the oldest known individual trees are California's bristlecones, giant sequoias, and western junipers. The oldest known living bristlecone, called Methuselah, germinated back when humanity was inventing writing, about 5,000 years ago.

read into the past like the pages of a book . . .

Tree rings let us read into the past like pages of a book, revealing

to the trained eye rich descriptions of past forests and their experi-

ence of fire, rainfall, temperature, their times of plenty and of intense

struggle. Their stories tell us about the experience of people, too—

when the land flowed with milk and honey and when it didn't, clues

to why past societies fought, moved, and sometimes disappeared.

It seems that trees don't grow old in the usual sense of progressive disruption of metabolic processes of renewal. Amazingly, trees' life processes might possibly go on forever in the absence of catastrophic events like fire or outbreaks of pests and disease, the primary nonhuman causes of tree mortality.

. . . trees' life processes might possibly go on

Long-dead trees can live on in beautiful ways. People employ exceptional grain—the lines revealed when water-bearing vessels are cut across—to add beauty in all manner of building and design. The prized grain of *Acacia koa*, a tree growing naturally only in Hawai'i, shimmers in many tones that seem to move in three dimensions.

forever . . .

Trees live on through sound as well. People have long debated whether and why the instruments produced by master violin makers of the late seventeenth and early eighteenth centuries have superior tonal qualities compared to those of today. Antonio Stradivari, the most celebrated violin maker, produced instruments from trees grown in the middle of the Little Ice Age, a period of longer winters and cooler summers—and thus used wood with slower, more even growth. These climate conditions, combined with the exceptional growing conditions in the southern Italian Alps where the wood was taken from, have not recurred since Stradivari's "golden period."

As Earth's forests change, so do her chemistry and climate. The Little Ice Age (ca. 1550–1850) may have been triggered in part by the regrowth of forest on abandoned farmland, following the arrival of Europeans in the Americas and the resulting decimation of native peoples. Since the Industrial Revolution, a substantial portion of the increase in the carbon dioxide content of the atmosphere now warming the planet results from deforestation by human hands. Today many forests are again recovering and are absorbing 25 percent of carbon dioxide emissions from human activity.

Trees live on through sound . . .

Trees have shaped human beings in profound ways. Living in trees for 80 million years or so, our ancestors acquired exceptional hand-eye co-ordination and dexterity. When the climate dried and they came down into the savannah—on their way to becoming uniquely human, walking upright and with hands free—they were primed to do the most human of things: to invent.

Trees fueled human discovery, literally. In the early days, after our African ancestors first mastered fire, burning wood unlocked cold northern environments and rich food resources inaccessible before cooking. By supplying the wood for primitive stoves, trees may have powered the evolution of large, calorie-hungry hominid brains. Humanity's unique social system is thought to have evolved around the earliest of those hearths. Wood propelled societies from the Stone Age into the Bronze Age and the Iron Age, when ores were smelted for the production of tools, weapons, and artifacts that marked increasingly complex cultures. Later the Age of Discovery was launched with fine-timbered ships that opened vast frontiers of knowledge and global exchange.

In modern times, people are just beginning to appreciate the many benefits of trees. Some insights seem small—like the recent finding that tropical rainforest can boost coffee yield by 20 percent and its quality by 27 percent. These boosts are thanks to myriad bees that live in the forest and pollinate crops on nearby farms when they bloom. (Most of the bees do not sting, though some, like the memorable *Scaptotrigona mexicana,* will yank out a person's eyebrow hairs when annoyed.) Small insights, yet 70 percent of the crop varieties that feed humanity get a yield boost from pollinators, an unknown fraction of which need forests to flourish.

Other benefits, harder to quantify but widely shown, are the physical, emotional, and cognitive boosts from experiences in nature. A famous study found that patients with trees outside their hospital windows recovered from surgery more quickly than patients whose windows looked out on brick walls. Trees' starring role in producing these benefits is evident in many societies, in sacred groves honoring natural spirits and in the designation and design of places for recovery of the human body and spirit. Whether living as great wild expanses or as ribbons and dots of connection and texture in human landscapes, trees define our lives and the future of humanity.

. . . trees define our lives

Acknowledgments

The images in this book were all made in the Skagit River region in Washington State, a landscape remarkable for its natural beauty. The Skagit River's 150-mile run originates in British Columbia, flowing from the mountains through forests, towns, and farms to its mouth in Puget Sound, 70 miles north of Seattle, and sustaining a watershed of 3,000 square miles. Charles Katz Jr. has known the lower Skagit area for much of his life and has long admired the way its people work together to protect its many natural dimensions while accommodating human needs. Gretchen Daily made two visits during the slow loss of her father. We wish to thank our friends at the Nature Conservancy of Washington for teaching us about the Skagit and for devoting their lives to that landscape and its people: Bob Carey, Roger Fuller, Danelle Heatwole, and the late David Weekes.

We are deeply grateful to colleagues and friends for their support and inspiration. We are indebted to many, but in particular we wish to thank Robert Adams, Peter and Helen Bing, Haydi and Damon Danielson, Paul and Anne Ehrlich, George Fisher, Jimmy and Emmy Greenwell, Elizabeth Hadly, Neil Hannahs, Ana Maria Herra, Peter Kareiva, Mona Kuhn, Patti Lord and John McMillan, Hal Mooney, Jim Salzman, Vicki and Roger Sant, John Schroeder, and Kelsey, Tim, and Wren Wirth. We are also tremendously grateful to associates of Stanford University's Center for Conservation Biology, including Bill Anderegg, Becca Goldman Benner, Larry Bond, Greg Bratman, Kate Brauman, Berry Brosi, Janet Elder, Luke Frishkoff, Josh Goldstein, Liba Pejchar Goldstein, Rachelle Gould, Danny Karp, Edith Katsnelson, Yicheng Liang, Chase Mendenhall, and Jai Ranganathan.

We give thanks beyond words to our families.

Published by Trinity University Press
San Antonio, Texas 78212

Trinity University Press strives to produce its books using meth-
ods and materials in an environmentally sensitive manner. We
favor working with manufacturers that practice sustainable man-
agement of all natural resources, produce paper using recycled
stock, and manage forests with the best possible practices for people, biodiversity, and sustainability. The press is a member of the
Green Press Initiative, a nonprofit program dedicated to supporting publishers in their efforts to reduce their impacts on endan-
gered forests, climate change, and forest-dependent communities.

The paper used in this publication meets the minimum
requirements of the American National Standard
for Information Sciences—Permanence of Paper for
Printed Library Materials, ansi 39.48-1992.

Book design and composition by Kristina Kachele Design, llc
This book is printed on 100lb. Sterling Matte
PRINTED IN CANADA

ENVIRONMENTAL BENEFITS STATEMENT

Trinity University Press saved the following
resources by printing the pages of this book on
chlorine free paper made with 10% post-consumer
waste.

TREES	WATER	ENERGY	SOLID WASTE	GREENHOUSE GASES
1	321	1	21	59
FULLY GROWN	GALLONS	MILLION BTUs	POUNDS	POUNDS

Environmental impact estimates were made using the Environmental Paper Network
Paper Calculator 3.2. For more information visit www.papercalculator.org.

Library of Congress Cataloging-in-Publication Data
Daily, Gretchen C.
The power of trees / text by Gretchen C. Daily;
photographs by Charles J. Katz Jr.
   p. cm.
Summary: "Conservation biologist Gretchen Daily nar-
rates the evolution, impact, and natural wonder of trees,
alongside 26 photographs by Charles Katz that illustrate
the development of trees: how trunks were formed, what
tree rings tell us about human societies, and how trees
define the future of humanity"—Provided by publisher.
ISBN 978-1-59534-132-7 (hardcover : alk. paper)
1. Trees—Evolution. 2. Trees—Ecology. 3. Trees—
Pictorial works. 4. Trees—Social aspects. I. Title.
QK477.D24 2012
582.16—dc23   2012010002
16 15 14 13   5 4 3 2

Gretchen Daily's work spans scientific research, teaching, public education, and advancing practical approaches to environmental challenges. Her scientific research is on countryside biogeography and the future dynamics of biodiversity change; the scope for harmonizing biodiversity conservation and agriculture; quantifying the production and value of ecosystem services across landscapes; and new policy and finance mechanisms for integrating conservation and human development in major decisions. Daily, who is Bing Professor of Environmental Sciences at Stanford University, cofounded and codirects the Natural Capital Project, an international partnership to mainstream the values of nature into major decisions of governments, businesses, and communities. Her honors include the Sophie Prize, the International Cosmos Prize, the Heinz Award, and the Midori Prize. She has published numerous scientific articles and books, including *The New Economy of Nature: The Quest to Make Conservation Profitable*, cowritten with Katherine Ellison.

Charles Katz Jr. has been an active photographer while pursuing a professional career as an attorney and business executive. He serves on the board of directors of the Nature Conservancy of Washington and on the boards of advisers for the Natural Capital Project, Stanford University's Woods Institute for the Environment, and Stanford's School of Earth Sciences. His previous publications in photography include *Etched in Stone: The Geology of City of Rocks National Reserve and Castle Rocks State Park, Idaho,* with text by Kevin R. Pogue.